IFLA

International Federation of Library Associations and Institutions
Fédération Internationale des Associations de Bibliothécaires et des Bibliothèques
Internationaler Verband der bibliothekarischen Vereine und Institutionen
Международная Федерация Библиотечных Ассоциаций и Учреждений

IFLA Publications 9

Standards for Public Libraries

Prepared by the
Section of Public Libraries

2nd Edition

Verlag Dokumentation, Publishers, München 1977

IFLA Publications
edited by
Willem R. H. Koops and Peter Havard-Williams

Recommended catalog entry:

Standards for Public Libraries. 2nd corrected Edition
München: Verlag Dokumentation, 1977, 53 p.; 21 cm
(IFLA Publications, 9)
ISBN 3-7940-4429-0

2nd corrected Edition

Publisher:
Verlag Dokumentation Saur KG, München
D-8000 München 71, POB 711009, Federal Republic of Germany

Printed and bound by Friedrich Pustet, Regensburg
ISBN 3-7940-4429-0

Guidance as to the levels of provision needed to maintain efficient public library services. It is recommended that these standards should be accepted as guidelines for the development of national public library services, and should provide a basis for the formulation of national standards.

CONTENTS

FOREWORD to the second edition

About four years after publication in 1973, the IFLA Standards for public libraries are out of print. It is evident that they proved their usefulness for the library policy and library practice in many countries of the world.

During these years there has been some further development specially in the use of audiovisual material in public libraries.

Here some countries worked out national standards. For the study of international standards and guidelines in this field the IFLA Section of Public Libraries established a working group on A.V. material. Awaiting the results of this study it was decided to refrain from revision of the relating paragraphs at this moment and to reprint the Standards with only some minor amendments.

The Hague, March 1977

Dr. P. J. van Swigchem,
Chairman Section of Public Libraries

INTRODUCTION

In 1956–1958, The Public Libraries Section of IFLA published two documents on Standards for Public Library Service, which were adopted by the IFLA Council. These standards were, in the terms of the document 'a statement setting out in simple terms the basic minimum standards for an effective public library service'. While it was recognized that social, economic and geographical conditions, and existing stages of library development varied so greatly that it was impossible to formulate standards in precise terms, it was considered that certain fundamental requirements could be stated in terms applicable to local and national circumstances.

The standards, as formulated, covered five principal aspects of public library service.

a. Books and other materials
b. Staff
c. Availability
d. Facilities and amenities
e. Library premises

The standards were never finalized, nor were they ever issued as a separate document, owing to the premature retirement through ill-health of the Chairman, L. R. McColvin. In spite of this, the standards have been widely quoted, and generally accepted either as standards complete in themselves or as a basis for consideration in many countries formulating standards.

Although the standards are still valuable, and many of the formulae put forward are still valid, the Public Libraries Section of IFLA in 1965 considered that in view of the many developments that had taken place the time had come for consideration of revised standards with some consideration of standards for special groups.

After the collection of information, in 1969 a letter was sent to all member Associations, setting out the intention to revise the standards and asking for comment. In particular, it was asked whether it was considered desirable that separate standards should be formulated for developing countries.

There was general support for the main proposal. Separate standards were *not* considered desirable, since the general objectives in all countries were the same, the modifying factor being the pace at which development could take place.

Accordingly, at the Moscow meeting of IFLA in 1970, a report was submitted, and it was agreed that new standards should be prepared. A representative Working Party was formed for the task, and included in it were members of the Committee on Library Buildings of IFLA.

Several countries have in the past few years published standards relating to their own services, and these have been taken into consideration by the Working Party. The correspondents of appropriate sections have also been consulted. The text here submitted is the result of the deliberations of the Working Party. It was submitted to the meeting of the Section at Budapest in 1972, and approved, subject to minor modifications.

The members of the Working Party are:

F.M. Gardner	United Kingdom	Chairman
J. Torfs	Belgium	Secretary
Mrs. L. Bradshaw	United States of America	
Mrs. Aase Bredsdorff	Denmark	Representing Library Work with Children Sub-section.
G. Chandler	United Kingdom	Representing Intamel Sub-section.
K. C. Harrison	United Kingdom	
H. Hoffman	Federal Republic of Germany	
A. C. Jones	United Kingdom	
Miss M. Joy Lewis	United Kingdom	Representing Libraries in Hospitals Sub-section.
W. Mevissen	Federal Republic of Germany	
I. Papp	Hungary	
G. Ruckl	German Democratic Republic	
W. Schumann	German Democratic Republic	

C. H. Ray (Library work with Children Sub-section) and Mrs. J. M. Clarke (Libraries in Hospitals Sub-section) were brought in for discussion on standards for their sections.

UNESCO PUBLIC LIBRARY MANIFESTO

Because of its relevance to desirable levels of provision, the UNESCO Public Library Manifesto, as revised in 1972, is set out in full as a preamble to this report.

UNESCO and public libraries

The United Nations Educational, Scientific and Cultural Organisation was founded to promote peace and spiritual welfare by working through the minds of men and women.

This Manifesto proclaims UNESCO's belief in the Public Library as a living force for education, culture and information and as an essential agent for the fostering of peace and understanding between people and between nations.

The public library

A democratic institution for education, culture and information

The Public Library is a practical demonstration of democracy's faith in universal education as a continuing and lifelong process, in the appreciation of the achievement of humanity in knowledge and culture.

The Public Library is the principal means whereby the record of man's thoughts and ideas and the expression of his creative imagination are made freely available to all.

The Public Library is concerned with the refreshment of man's spirit by the provision of books for relaxation and pleasure, with assistance to the student and with provision of up-to-date technical, scientific and sociological information.

The Public Library should be established under the clear mandate of law, so framed as to ensure nation-wide provision of public library service. Organized co-operation between libraries is essential so that total national resources should be fully used and be at the service of any reader.

It should be maintained wholly from public funds, and no direct charge should be made to anyone for its services.

To fulfil its purposes, the Public Library must be readily accessible, and its doors open for free and equal use by all members of the community regardless of race, colour, nationality, age, sex, religion, language, status or educational attainment.

Resources and services

The Public Library must offer to adults and children the opportunity to keep in touch with their times, to educate themselves continuously and keep abreast of progress in the sciences and arts.

Its contents should be a living demonstration of the evolution of knowledge and culture, constantly reviewed, kept up-to-date and attractively presented. In this way it will help people form their own opinions, and develop their creative and critical capacities and powers of appreciation. The Public Library is concerned with the communication of information and ideas, whatever the form in which these may be expressed.

Since the printed word has been for centuries the accepted medium for the communication of knowledge, ideas and information, books, periodicals and newspapers remain the most important resources of public libraries.

But science has created new forms of record and these will become an increasing part of the Public Library's stock, including print in reduced form for compact storage and transport, films, slides, gramophone records, audio and video tape, for adults and children, with the necessary equipment for individual use and for cultural activities.

The total collection should include material on all subjects, to satisfy all tastes at differing educational and cultural standards.

All languages used by a community should be represented, and there should be books of world importance in their original languages.

The Public Library building should be centrally situated, accessible to the physically handicapped, and open at times convenient to the user. The building and its furnishings should be attractive, informal and welcoming and direct access by readers to the shelves is essential.

The Public Library is a natural cultural centre for the community, bringing together as it does people of similar interests. Space and equipment are therefore necessary for exhibitions, discussions, lectures, musical performances and films, both for adults and children.

Branch libraries and mobile libraries should be provided in rural and suburban areas.

Trained and competent staff in adequate numbers are vital to select and organize resources and assist users. Special training will be required for many activities such as work with children and the handicapped, audio-visual materials, and the organization of cultural activities.

Use by children

It is in early life that a taste for books and the habit of using libraries and their resources are most easily acquired. The Public Library has therefore a particular duty to provide opportunity for the informal and individual choice of books and other material by children. Special collections and if possible separate areas should be provided for them. The children's library can then become a lively stimulating place, in which activities of various kinds will be a source of cultural inspiration.

Use by students

Students of all ages must be able to rely on the Public Library to supplement the facilities provided by their academic institutions. Those pursuing individual studies may be entirely dependent on the Public Library to meet their needs for books and information.

The handicapped reader

There is an increasing concern with the welfare of the elderly and all handicapped people. Problems of loneliness, and mental and physical handicaps of all kinds can be alleviated in many ways by the Public Library.

Improved means of access, provision of mechanical reading aids, books in large print and recorded on tape, service in hospitals and institutions, and personal service to the home are some of the ways in which the Public Library can extend its services to those who need it most.

The Public Library in the community

The Public Library should be active and positive in its outlook, demonstrating the value of its services and encouraging their use.

It should link itself with other educational, social and cultural institutions, including schools, adult education groups, leisure activity groups and with those concerned with the promotion of the arts.

It should be watchful for the emergence of new needs and interests in the community, such as the establishment of groups with special reading requirements, and new leisure interests to be represented in the library's collections and activities.

SOME GENERAL PRINCIPLES, BASED ON THE UNESCO MANIFESTO

i There should be generally applicable and mandatory legislation.

ii There should be a central authority charged with supervising the implementation of the legislation.

iii The legislation, national or state (in a federal state), should name the appropriate units to be charged with responsibility for providing public library service. These units should be the largest possible which can directly administer a comprehensive and efficient service. Smaller units may be given a measure of responsibility, subject to supervision.

iv The cost of the public library service should be met from public funds provided by central or local government, or by a combination of both. In developing countries in particular, there must be substantial support from central government. There must be provision for continuing funds, and all local authorities with some degree of responsibility for supervision should have power by legislation to raise funds.

v Use of the service should be free of direct charge to the user, except when material becomes the property of the user, e.g. photocopies, catalogues.

vi It must be recognized that no single administrative unit of public library service can be self-supporting in supply of materials for its users. Legislation should therefore provide administrative machinery for the overall planning of library and information services, co-operation between library administrative units, and between public libraries and other libraries. There should also be provision for the setting up of centralized services, e.g. cataloguing, classification, binding, indexes and bibliographies, and research.

vii It should be clearly laid down that the only public library service which can be supported by public funds is that provided under law, and provision should be made for the amalgamation of all public libraries receiving financial support from public funds into the general public library service.

viii The requirements of users both for borrowing and consultation should be studied and provided for, both in a general and local context. Attention should particularly be given to local requirements for reference material for study and information, and for current periodicals and newspapers. Assistance should be given to users by members of the staff and through the provision of catalogues, booklists and appropriate bibliographies.

ix Despite the recognized importance of the public library in cultural and educational development, it must, if it is to fulfil its purposes, maintain its identity, and remain an independent institution. This need in no way hampers the library's sponsorship and encouragement of cultural projects, nor prevents association at a local level with community organizations, social education, or literacy promoting organizations, particularly in developing countries, always supposing that the smaller libraries or service points are organizationally and closely linked to the central library of a public library system.

x It should be recognized that in all countries, the full potentiality of a public library service cannot be realized without professionally trained and qualified staff. This implies training facilities, recognition of librarianship as a profession needing higher education and a nationally recognized scale of salaries.

xi These standards should be accepted as guidelines for the development of national public library services, or should provide a basis for the formulation of national standards.

THE NEED FOR STANDARDS

1. A public library, by its nature, must be a continuing and growing organization. There must be finance to house it, to purchase its materials and equipment, and to pay its staff. Printed materials must be available in all languages used by the community, and for all sections of the community, and these resources will be increasingly amplified by audio-visual materials. It must also be possible to recruit trained or trainable staff.

2. A public library service should not be created unless there are financial resources to start and support it and materials and staff to maintain it. It is considered, however, that there are few countries today which cannot meet these requirements.

3. It must be expected that the emphasis placed on the various objectives of the service – and the means of achieving these objectives – will be different in different communities, and will change with the passage of time. Thus, we have seen a general increase in the importance of library provision for children, a sharp rise in the provision and use of technical information, and greater use of audio-visual materials of all kinds. The emphasis placed on various aspects of library provision will be quite different in a small rural community from that in an industrial town; in a developing country emphasis may need to be placed on simple educational material and books for leisure reading, changing later to more sophisticated provision, including technical and commercial information.

4. In this fluid situation it is unrealistic to seek standards which will have universal validity. The most that can be offered are guidelines based on past experience in communities where circumstances are bound to be different from those in the communities where guidance is needed.

5. The guidelines in this document are based, in the main, on the experience of librarians in countries with developed library services. Librarians elsewhere will find it useful to know the types and levels of provision which have been found generally appropriate in this situation. No doubt many of these will be applicable elsewhere; but librarians who are working in other circumstances, with different priorities for their services, must themselves determine the extent to which these guidelines can usefully be followed.

6. Standards are inter-dependent; for instance, the rate of acquisition, cataloguing and classification of materials will depend on the rate at which trained staff can be provided. Similar levels and types of service may not be reached, and may not be needed, over a whole country at any one time. But though progress may not be uniform, it should not be allowed to vary too widely. Consistent standards and the collection of truly comparable statistics will allow progress to be measured, so that emphasis can be changed if necessary.

7. Levels of provision must be related to the needs of individual users, because a public library does its work with individuals. The efficiency of a library service will be assessed in terms of what an individual can obtain, whether through a postal service, a mobile library, a branch library or a large city library. Manifestly, accessibility is a major factor in what service can be given, but small service points need to be linked to larger units if the isolated user is to be given the best possible service.

UNITS OF ADMINISTRATION AND SERVICE

Some definitions

The following terms are used throughout this document:

8. *Administrative unit* – A unit of public library service which is independent in the sense that it has its own governing body, receives and manages its own funds, and appoints its own staff.

9. *Public Library System* – A group of public libraries and services forming an independent administrative unit, *or* comprising a number of such units acting together by agreement.

10. *Service point* – A part of an administrative unit or library system from which direct service is given, e.g. a deposit collection, mobile library, branch library, main library, children's library or hospital library.

11. *Public Library Service* – A general term expressing function, not area or administrative control.

The size of the administrative unit

12. It is increasingly recognized that in developed, urbanized communities, very large units, in terms of population, are necessary to support comprehensive library services, and the third of the general principles already enunciated states that 'these units should be the largest possible which can directly administer a comprehensive and efficient service'.

13. There are, however, many factors such as economic resources and density of population, which may prohibit the creation of ideal administrative units. Local government systems also vary widely from country to country: a densely populated and economically viable area may be fragmented into many independent local authorities, none of which is capable of providing a comprehensive public library service, but which are unable or unwilling to transfer their powers to a larger authority (if one exists). Lack of communication may be another factor making the co-ordination of services impracticable over a wide area.

14. Nevertheless there is evidence of a definite tendency, shown by recent legislation or executive action, towards the creation of larger administrative

units for library purposes. This may result from changes in local government systems, transfer of responsibility to larger local authority units, creation of 'regional' systems, or by contracts between existing administrative units. Even the largest public library system, however, cannot be totally self-sufficient in view of the range of materials and services required, and co-operation between systems is therefore essential. It should be noted that INTAMEL has prepared standards for library services in cities of over 400,000 population.

15. A public library system should preferably have a population base of at least 150,000. It is only in this situation that it becomes economically possible to provide a comprehensive range of materials, to make use of sophisticated modern techniques and equipment, and to provide a full range of specialized services with appropriately trained staff. Although the administrative unit will usually be a single local government unit, experience has shown that the creation of library systems by linking independent units of local government is perfectly feasible.

16. In some countries library services have been provided on a national basis by the central government. This method may be especially appropriate in a developing country where a large administrative unit is particularly important if limited resources are to be used to the greatest effect.

17. Large administrative units make possible a faster rate of library development and better standards of service, since the requirements of general users can be met without neglecting more specialized needs. However, it is not possible in all communities to provide such large administrative units. A population of 50,000 should normally be considered the minimum acceptable size. In exceptional circumstances it may be necessary for much smaller communities to undertake provision of their own library services.

18. The smallest administrative unit for public library purposes that can be contemplated in any circumstances would have a population of 3,000. Below this figure no useful standard can be defined. It should be made absolutely clear that such a figure is only acceptable if no other solution is possible – e.g., in areas with an extremely low density of population, or extreme difficulty of communications. Wherever possible, the administrative unit should be a much larger one in terms of population. The smaller the unit, the more important it is that it should co-operate with other units.

Service points: types and use

19. In a developed library service, urban main libraries should be open for a minimum of 60 hours a week. In branch libraries, opening hours may vary from 18 to 60 hours a week according to need. Where purpose-designed libraries are open for less than 40 hours a week, some arrangement for using the accommodation outside library hours – possibly for cultural activities, will be desirable if they are not to be seriously uneconomic to run. All libraries should be available at times convenient to users, usually including evenings and weekends, and should be easily accessible to all members of the community including the elderly and the physically handicapped. Even the smallest administrative unit, of 3,000 population, will require a library housed in a separate building, or in adequate permanent space in a larger and suitable building, possibly a community centre or school.

20. Main libraries and large branch libraries provide a service in depth to a population drawn from a wider area, smaller branch libraries and other service points provide a more limited service to their local population, many of whom look to the larger libraries to satisfy their more searching demands.

21. Very small communities and isolated homes need to be served by mobile libraries, deposit collections or postal services. Where mobile service is possible the aim should be to visit each stopping place at least once a fortnight, the pattern of stops being determined by local conditions. Mobile libraries need to be able to draw on an extensive bookstock, and should therefore be based on the strongest possible fixed service points. Deposit collections, sometimes the only practicable alternative to mobile libraries, are expensive in regard to bookstocks, and are often frustrating to users. They are acceptable only when difficulties of communication leave no alternative. A combination of mobile service and smaller service points with a standing core collection on the premises may be considered for smaller settlements.

STANDARDS FOR COLLECTIONS

Bookstocks

General

22. In the smallest administrative units, at least three volumes per inhabitant are required, but this requirement will lessen as population rises, and a satisfactory general standard is two volumes per inhabitant. These standards should allow for approximately one third of the total stock to be for children, where children up to 14 years of age constitute 25%–30% of the population. Where the proportion of children is higher than 30% the provision will need to be correspondingly higher.

23. The above recommendations are for a minimum working stock, with regular discard of older and worn-out material. They cover all departments and purposes – reference and lending, children and adults. They would however exclude collections for special groups of users, such as handicapped readers and non-indigenous groups.

24. In countries with developing library services, the recommended levels of provision cannot be attained immediately, but in small communities in particular, the stock of books should at least be in sight of the standards laid down. A stock of less than 9000 volumes cannot give a range of materials, nor be considered an adequate base for satisfactory service.

25. Generally, the standards relate to literate population, but even so the stock of 9000, appropriate to a unit of 3,000 population, is considered a minimum. With a well-maintained stock of 9000 volumes, a public library service becomes at least a viable project, and it could even from its own resources give some help to smaller communities or isolated users within reach.

26. Within any administrative unit, communities smaller than 3,000 will require special services, by means of mobile libraries, deposit collections or postal loans. Deposit collections should be changed not less than four times each year, at least 200 volumes being included in each delivery. A few reference books should be permanently available, and a catalogue or booklists should be provided so that users can order particular books in advance, or by post between deliveries.

Reference books

27. For the purpose of these standards, a reference book is one that must be available at all times, and so cannot be loaned. The average price of reference books is relatively high, and many of them need to be replaced regularly. Even in a small community, a basic collection of reference material is necessary, and this is particularly true in developing countries, where there may be few reliable sources of information outside the library. Provision may, however, be limited by the insufficient production of reference works relevant to the country in question, or in indigenous languages.

28. In the smallest administrative unit of 3,000 population, the minimum collection should be at least 100 reference titles. In larger communities, reference material may need to constitute up to 10% of total stock. In very large administrative units the proportion may well be even higher. In view of the heavy cost of maintaining reference collections, consideration should always be given to the possibility of co-operating with other national and local libraries.

29. The special needs of children for reference material should be considered, even in branch libraries. While older children should be given access to adult reference collections, separate provision of simpler material should be made available in the children's library.

Withdrawal of material

30. Books need to be withdrawn if they have become physically worn out, if their contents are dated or if their value to a particular community has been exhausted. Good physical condition alone is not sufficient reason for retaining a book on a library's shelves, but the useful life of some books will be extended by transfer to other libraries.

31. In a new library, withdrawal through physical wear will obviously not be an important consideration at first. In smaller libraries also, the proportion of stock to population is high, so physical wear will not be so apparent as in large libraries. In large libraries, however, the number of books is larger in total but smaller in relation to population, and thus physical wear becomes a more important factor, since more readers will be using proportionately fewer books. The proportion of funds required for replacement will therefore tend to rise as a library becomes established and stocks grow.

32. In developing countries, where books are not produced to a high standard of paper and binding, physical wear is a very serious problem, and this must be allowed for in the provision for withdrawal and replacement of worn-out stock.

Annual additions

33. These include newly published books, including children's books and adult fiction, replacement of out-of-date material, particularly technical books, and books on new subjects. The maintenance of a balanced stock is important, and the object should be not merely to maintain, but to improve the stock. In a sense, the provision of a higher proportion of new books in a small stock is more important than in a large one. In a small library to a reader with special interests, such as horticulture, the number of books available will be small, and the stock may therefore appear to be static.

34. It is recommended that additions should be at least 250 volumes per annum per 1000 of population in each administrative unit, but in some countries, and particularly in smaller units, 300 volumes would be needed to provide a reasonable proportion of titles to cover the great variety of subjects on which material is available.

35. Unless the rate of use is exceptionally high, this level of provision will maintain stock in adequate physical condition, provide a reasonable intake of new books, and enable regular reviews of out-of-date material. Approximately one-third of these additions should be for children, where the child population is 25%–30%. Where the higher standard is approached, most of the additional allowance will be for specialized nonfiction. Reference books will constitute only a small proportion of annual additions in very small libraries, but the proportion will tend to increase with size until, in libraries serving populations of over about 50,000 they may account for 10% of additions.

36. The standard for accessions may sometimes have to be modified to take account of the amount of national book production, which may be unbalanced in relation to public library requirements. This applies particularly in some developing countries. It will usually be necessary however to supplement the country's own output by acquiring books published in other countries and in other languages, particularly in technical subjects.
A sufficient amount of electrical outlets is necessary in order to take care of the new technological machinery.

Binding

37. Books purchased in paper covers will have a longer life if they are bound – or laminated – before use. Many books purchased in publishers' bindings will need to be rebound later in their lives, before they become unusable. Regular review of the need for binding, and of the funds necessary for this purpose is therefore of the utmost importance.

Periodicals, including newspapers

38. In developed countries there are great variations in the provision of periodicals and newspapers in libraries. and in some countries standards have not risen as have standards for the provision of books. Even within countries, provision varies widely between administrative units. Since periodicals are a vital documentation resource, a standard for guidance is greatly needed.

39. A basic provision of at least 50 periodicals is considered necessary, and in larger administrative units 10 periodicals per 1000 population should be provided. This would include multiple copies for different service points, periodicals in foreign languages, and periodicals for children.

40. In administrative units of 100,000 population or more, the scale of provision needed will depend upon such factors as the availability of subject indexes, and the extent to which co-operation in the provision of periodicals has been organized. It is likely that in very large communities the scale of provision related to population can be reduced, but it is not possible to make any recommendation as to the scale appropriate in different circumstances.

41. The provision in libraries of periodicals (including newspapers) may be particularly important in developing countries with low standards of literacy, since periodicals are often especially valuable for the newly literate. In such countries, moreover, the production of periodicals is usually in advance of book production.

Audio-visual materials

General

42. Audio-visual materials are understood here to be those which require equipment for listening or for viewing. They include gramophone records, magnetic tapes, films, slides, filmstrips and video recordings. The use of these materials has developed only recently and they are still not commonly available in all countries. Nevertheless their importance in libraries is now increasing rapidly.

43. In general, the provision of printed material should have priority over the provision of audio-visual materials. In developing countries in particular the cost of providing audio-visual materials, and the equipment needed for its use, must be weighed very carefully against the potential use of the material. A greater volume of use can probably be expected to result from expenditure on printed material than from a similar expenditure on audio-visual material, but audio-visual materials may provide a more effective medium for certain kinds of communication. There is no doubt, however, that the potential use of this material is very great.

44. Closely allied to audio-visual materials are the many kinds of microform material which provide a convenient means of storing information which is particularly bulky in its printed form, or a relatively cheap means of acquiring material which in printed form is unobtainable. Although a vast quantity of microform material is now available, its relevance to the needs of different countries varies considerably, so that no generally applicable standards of provision can be recommended. It should be noted, however, that public libraries can themselves influence the provision of material in this form, and in some cases can themselves take a lead by transferring non-copyright material to microform.

45. All public library services should have the following equipment readily available for use in both adult and children's departments: record player, tape recorder, cine projector, slide or film-strip projector, cassette player, portable screen and radio and television receivers.

Gramophone records and magnetic tapes

46. A minimum collection to be reasonably adequate for a population of up to 20,000 should consist of not less than 2000 records and/or tapes,

including not only music, but also speech records, instructional records and language courses, both for adults and for children.

To maintain such a collection, at least 300 records should be purchased annually, or more where home lending is permitted. Withdrawal of badly worn or out-of-date recordings is an important factor in the maintenance of this collection.

Films, slides, film strips and video recordings

47. All libraries should consider provision of these materials, though no particular level of provision can be generally recommended. Colour slides and transparencies are now plentiful, and could be made available for loan, since many people own viewing equipment. Film strips are also available in wide variety, though their use is more usually limited to educational and instructional purposes.

Art collections and other materials

48. Many libraries provide for reference, and sometimes for loan, collections of reproductions of paintings, prints or even of original art works. These special art collections, though falling outside the category of audio-visual materials as used here have their place in multi-media public libraries. No standards however can be recommended for their provision.

STANDARDS FOR SPECIAL GROUPS

Children

49. Children form an identifiable group within the community having special needs and interests and for this reason specific library provision must be made to cover the child's need at all ages and ability levels from infancy to the age of 14 years. It is important, however, that such provision should be integrated physically and administratively with the total public library service: all children at some time, and some children for much of their time, will need to draw on wider resources than those of the children's library. As far as possible therefore the standards for children's provision have been integrated with those for adults under the relevant heading.

50. The role of the library in a child's development is not limited to the provision of factual information; the library also has great importance in stimulating the imagination. For this reason, a balance must be kept between fiction of all kinds and non-fiction stock. The range of representation should take into account the needs of children, not only of school age, but at pre-school and pre-reading levels; in consultation with local educational services, provision is also needed for children of low ability; and the needs of parents and others concerned with children should be met. While the demand on children's libraries may be especially heavy where adequate school libraries do not exist, the standards refer only to the provision for children through the public library; there is a need for separate school library provision, organized where appropriate in co-ordination with the public library service.

51. Audio-visual materials, including those produced principally for educational purposes, are of particular value in the field of children's work both within the library and for external use. Attention is drawn to the possibility of using filmstrip and slides as a means of exploiting books, and to the importance of gramophone records and magnetic tapes in the fields of the spoken word and of music.

52. Staff in children's libraries need special knowledge of the library needs of children and the resources available to meet them and must be involved both with schools and with other organisations and institutions (such as youth groups, pre-school institutions, and parents' groups) and

persons (such as teachers and those responsible for their training) operating in the same field.

Handicapped readers, including the housebound and those in hospitals or correctional institutions

General

53. All public library services should be available and accessible as far as possible to handicapped and housebound readers, both adults and children. The term 'handicapped' includes the mentally handicapped (mentally subnormal), the mentally ill (psychiatric) and the physically handicapped. The library should also aim to provide special services to them, and to hospitals, residential homes and correctional institutions within its area. These institutions will contain many people who have not hitherto appreciated the value of using public library facilities.

54. Where there is a considerable need for services of this kind, a separate department should be established, based on, and housed at, the public library, with a qualified librarian in charge and supporting staff as required. Although this service must be able to draw upon the total public library resources including periodicals, audio-visual materials and items specially requested, it is desirable to have a separate basic stock for the service including such special items as large print material where available. Reading aids, such as page turners and book holders, should be provided from a central collection.

Housebound readers

55. Home delivery services to housebound readers are expanding rapidly. With changes in medical and health care there is a general trend towards care in the community rather than a lengthy stay in hospital for many people. Local conditions will determine whether delivery services to the housebound can be given through branch and mobile libraries, or whether they should be centrally organized.

56. Visits to individuals at home should be made not less frequently than once in three weeks. The initial visit should be made by a qualified librarian although unqualified library staff or volunteers may be used for some of the visiting. In the most favourable circumstances (in a compact area where transport is available) it should be possible for one staff member to visit

20–25 persons in individual homes in a day. A library van may be required specifically for this purpose. There should be no limit to the number of books lent to any one reader and it may be convenient to pre-select some of the titles according to the known requirements of individual readers.

The visually handicapped, including the blind

57. In most countries there are separate national agencies providing library services for the blind. The existence of these should be known to all library staff and public libraries should provide information about such facilities. In some countries public libraries may themselves be agencies for the distribution of talking books and embossed literature. They also have a part to play in meeting the needs of the visually handicapped by providing and publicizing large print material.

Hospitals

58. A hospital with 500 beds, and a staff probably between 700 and 1000, needs a general bookstock of 4,000–5,000 volumes, regularly refreshed and supplemented by other material as noted in paragraph 54. The smallest hospital unit needs deposit collections of at least 200–250 volumes changed not less than four times a year. It is assumed that these stocks will be available to both patients and staff; the latter will also need their own professional library and information service. Detailed standards for hospital libraries have already been approved by IFLA and published.

59. The public library and the hospital, in co-operation, should ensure that suitable premises for the library service are available. Even very small hospitals will require accommodation for the storage of books, and there will usually be a need for a library room which can be visited by ambulant patients and staff. In all hospitals there should be a library trolley service to wards at least once a week. Long-stay hospitals, where extension services are provided will require additional space. At least one full-time librarian is essential in a hospital of 500 beds, and in special hospitals with perhaps fewer beds e.g. neurological, orthopaedic, psychiatric. Additional staff, including clerical personnel, will be required as a service develops.

Day centres, clubs and residential homes for the elderly

60. Deposit collections should be supplied to all such establishments on the scale of 2–6 books per head with an absolute minimum of 200 volumes, changed at least four times a year. A qualified librarian should visit them at least four times a year and additionally when requested. These establishments could sometimes be serviced by branch or mobile libraries.

Prisons, remand homes and detention centres

61. The public library should offer a library service, or a deposit collection, to all such establishments within the area. In general it is necessary to provide 5–10 volumes per head of library users within the institution, the number depending upon such factors as the average period of detention, the opportunities of access to the collection and the role of the library within the institution. It is desirable that a proportion of the stock should be changed at least four times a year and the service should be able to draw on the full resources of the public library to meet specific requests.

Library accommodation should be provided in such a way that the minimum restrictions need to be placed upon access by prisoners. A qualified librarian should visit all institutions in this category regularly, preferably at least four times a year, and additionally when requested. In large establishments a full-time librarian is needed.

National minorities and non-indigenous groups

62. A distinction must be made between permanent resident national minorities and non-indigenous immigrant groups. Permanent resident national minorities should be provided with literature in their mother tongue on the same terms as the general standards, even though this also is available on a scale proportionate to the need.

63. The provision of a service to non-indigenous groups now concerns many public library services. Special efforts should be made to encourage non-indigenous groups to use the public library. Arrangements may be needed on a national scale to solve the problems of meeting their needs, which may include the difficulty of obtaining books and periodicals in their native languages, and the fact that their stay in a particular community may be of limited duration.

64. Any identifiable non-indigenous group of 500 or more people in a community should be recognized as requiring public library service in their own language, including both books and periodicals. Collections should be on the scale of 1 book per 5 people up to 2,000 population and on the scale of 1 per 10 people beyond that. The minimum collection should be 100 books. Annual purchases to maintain and supplement these stocks should be at least 1 book per 25 people up to 2,000 population, and 1 per 50 beyond that. As noted in paragraph 23, these books should be in addition to the basic stock of 2–3 volumes per head of population.

65. For every 500 people there should be at least one periodical or newspaper in their own language, additional to the periodicals provided by the general standards. For smaller groups, periodicals may be provided from centralized collections, or by co-operation between libraries.

STANDARDS FOR STAFF

66. The staff referred to in this section are the professional, administrative and clerical staff employed in libraries. Manual workers such as cleaners and drivers – unless these also perform clerical duties – are not included.

67. Staff requirements for public libraries must be related to three factors: the population of the community served; the volume of use; and the range of services provided. The number of users and the number of books used will tend to increase rapidly in a developing service. At a later stage, when the amount of use measured by these means is expanding more slowly, a marked increase can be expected in demands on staff for information and assistance.

68. Population is regarded as the most satisfactory basis for the formulation of staffing standards, particularly as far as qualified staff are concerned, but the other factors mentioned above may require the modification of any standard based on population alone: this will be particularly true in a developing service, where the value to be attached to all three factors needs to be regularly reviewed. Special staffing standards will be necessary where extensive mobile library services are in operation.

69. In any public library service, qualified librarians are required to administer and supervise the service, to select and classify books and other materials, to undertake professional and technical assistance to readers, maintain contact with educational, cultural and community organisations and to plan the development of the service.

70. Clerical staff are required to carry out routine duties such as maintenance of records, recording and discharging loans, and preparing and maintaining collections. Persons qualified in other fields will also be required in other areas of the developed library service.

71. In the postulated minimum administrative unit of 3000 population with 9,000 books, some periodicals and a service both to adults and children, a full-time qualified librarian will be justified. Some clerical assistance will also be necessary.

72. For larger administrative units with developed library services, a total non-manual staff, including qualified librarians and others, amounting to one

per 2,000 population will normally be required. The need may be somewhat less in very large administrative units, but even in units of 150,000 population upwards, the requirement will be at least 1:2500 population. In large communities, however, the volume of administrative and clerical work becomes an important factor, and one would also expect there to be special services requiring additional qualified and trainee staff not allowed for in the above formula. Where there are large national minorities or non-indigenous groups these will require additional staff able to speak and read the relevant languages. In the future it is likely that a wider range of specialists – including administrators, technicians and experts in data processing – will need to be employed in most large libraries.

73. Where an administrative unit is giving a full range of services at least one qualified librarian will be required at each main service point, including the separate departments of main libraries and large branch libraries. Even the smallest service points of all library systems need someone permanently in charge, with some basic training in the purposes and methods of the library service, and under the general supervision of a qualified librarian.

74. The proportion of qualified librarians to total staff will depend on the specific conditions applying to an administrative unit. The minimum standard proposed for a developed urban, compact unit is 33% of total staff. In a unit with many branches and small service points, 40% would be a more realistic figure. In a service point serving a population of 10,000, one of the professional librarians should be a specialist in children's work. In larger library systems about one-third of the professional librarians will need to have a specialist interest in this aspect of librarianship.

75. The cost of staff is a very high proportion of the public library budget, and it is important that the salaries paid to qualified librarians should be sufficient to attract and adequately reward persons of ability and initiative. There should therefore be a clear division of work between qualified librarians and clerical staff and efficient management should ensure that the time of qualified librarians is not wasted on routine duties of a clerical nature.

76. In developing countries, lack of training facilities may make the appointment of sufficient qualified librarians difficult in the early stages. Initially, these librarians may, in some cases, have to obtain their training in

other countries. Where this is the case it is particularly important that the developing public library service should organize in-training facilities, until training within the educational system of the country becomes feasible.

STANDARDS FOR BUILDINGS

Considerations affecting planning

77. Although it will usually be necessary to provide in each public library building, lending and reference facilities for both adults and children, the relative importance of these will vary considerably in different circumstances. A decision must be made about the range and extent of these and other facilities, depending upon locally determined objectives and priorities, before the space required for each can be estimated.

Flexibility

78. All public library buildings should be planned to meet the needs of the next ten to twenty years so far as these can be anticipated, with due allowance for changes in the numbers and characteristics of the population served. If possible, the site should allow the future expansion of the building, and the plan should make this feasible. Within the building the plan – including arrangements for lighting, heating and other technical services – should allow the greatest possible flexibility in the use of space, especially in those parts open to the public. Even in large libraries it will often be better to provide *areas* for different purposes, separated where necessary by furniture or light partitions, rather than separate rooms or departments. Moreover it will be advisable, especially in larger libraries, to build up collections of foreign reference works and world classics in their original language.

Siting and accessibility

79. It is more important that libraries should be conveniently accessible than that they should have any specified radius of service. Libraries which are situated at focal points in the community, in the main stream of pedestrian traffic, well served by public transport and with ample car parking facilities, will attract more users, and users from a much wider area, than others which do not have these advantages. The relevant factor is the attractiveness of the focal centre as a whole, not that of the library alone, and this will usually point to the desirability of siting libraries in shopping centres. Large and efficient libraries will attract use from a wide area, and in general it can be expected that the larger and better the library, the greater will be its effective radius.

Inevitably, however, effectiveness decreases with distance and in urban areas it will usually be found necessary to provide branch libraries within 1 ½ km (1 mile) or so of most residents, and relatively large libraries within 3–4 km (2 – 2 ½ miles). These distances may need to be modified to take account of the transport facilities, topographical features, and the sense of community in different areas. The planning of the building itself should facilitate access by eliminating as far as possible steps, narrow doorways, and other confined spaces. The problems of the elderly and the handicapped should be kept in mind, by the provision of lifts, ramps and automatic doors where needed.

Association with other recreational, cultural and educational facilities

80. Provided that the above conditions can be met, there will often be advantage in linking the public library with other community services such as exhibition halls and theatres. In certain circumstances it may even be possible for a public library to serve simultaneously as the library of a school or other educational institution, especially when that institution aims to meet more comprehensively the cultural and social needs of the community. By such means as these the community will often be able to gain the maximum use and value from a number of related facilities.

Visual impact

81. The purpose of the building should be immediately apparent. The situation and design of a public library can achieve valuable publicity for the services which it provides. It will often be possible to provide views into the library to enable passers-by to see its public services in action, and exterior display cases and a well-planned entrance hall can be used to similar effect.

The standards

82. In the following paragraphs the space requirements for the different facilities provided by public libraries are considered one by one, together with the administrative and other areas which need to be associated with them. These recommendations relate only to public library facilities. When a library serves, for example, both the public and a school or college, considerable modification may be necessary.

83. The areas recommended for lending and reference services in paragraphs 86 to 92 relate to these functions *in toto.* Often the two services will be given from a single 'open-plan' room or department. Somtimes, however, part of one or both may be given from a separate department or separated area, concerned for example with music, technology, local history or a separate library for teenagers. When any area is separated in this way the total of the separated portion may need to be increased to provide for additional circulation space and to accommodate the additional use which the need for a separate department often implies – or which the provision of a separate department may sometimes stimulate. Children below 14 years of age usually constitute 25–30% of the total population, and the recommendations which follow assume this age pattern, though it is taken for granted that many children below the age of 14 will make considerable use of adult library facilities.

84. Attention is drawn to the overlapping functions of large and small service points within a single administrative unit. (See paragraph 20) An assessment of the population served will usually be the first step towards estimating the space requirements for each of the library services provided. It may often happen that a library which is the major reference library for a large area is also providing a lending service to adults in a somewhat smaller area and a service to children within its immediate neighbourhood. For each of these three functions it may therefore be necessary to estimate space requirements on different population bases.

85. It must also be noted that the general standard for bookstock discussed in paragraph 22 relates to administrative units as defined in paragraph 8. Within each administrative unit the bookstock will often need to be distributed among several service points, including main libraries or larger branch libraries whose service areas as explained above, overlap with those of the smaller branch libraries. It follows that, in these circumstances, the adult lending bookstock circulating from a particular service point at any one time may be less than that suggested in the following paragraph, and the figures suggested as a guide in Table 1 may need adjustment to take this into consideration.

Adult lending areas

86. Of the total bookstock of two to three volumes per head of total population, proposed in paragraph 22, it will usually be found necessary to

allocate not less than one volume per head of total population for the purpose of providing a lending service to adults (see Table 1). Of the total adult lending stock provided at each service point, it can be expected that approximately one third will be on loan at any one time, and this, together with the overlap between service points discussed in the preceding paragraph, suggests that as a general rule open access shelving should be provided in adult lending areas sufficient to accommodate 600 volumes per 1.000 population. Libraries serving populations larger than 60,000 or so will usually be able to provide somewhat fewer books per 1,000 population. The number of volumes displayed on open shelves in any adult lending library serving a population of 3,000 or more should never be less than 4,000. (See paragraph 25, which recommends a minimum total stock of 9,000 volumes, of which one third might be for children. This leaves 6,000 volumes for adults, most of which will be for loan, and about a third will be on loan at any one time). In small branch libraries, however, there may be a considerable seasonal variation in the number of books on loan, and to allow for this the floor area in adult lending departments should never be less than 100 m^2. (1076 sq.ft.), with shelf capacity in proportion.

87. In all larger libraries, allow 15 m^2. (161 sq. ft.) for every 1,000 volumes on open shelves. This assumes that shelf units for adults will be five shelves high. Provision on this scale will allow for circulation of readers

Table 1 Adult lending facilities

Population served (See paragraph 84)	Open-shelf accommodation Volumes per 1,000 population.	Total capacity	Floor area at 15 m^2 per 1,000 volumes (minimum 100 m^2)
3,000	1,333	4,000	100 m^2 (1076 sq.ft.)
5,000	800	4,000	100 m^2 (1076 sq.ft.)
10,000	600	6,000	100 m^2 (1076 sq.ft.)
20,000	600	12,000	180 m^2 (1938 sq.ft.)
40,000	600	24,000	360 m^2 (3875 sq.ft.)
60,000	600	36,000	540 m^2 (5813 sq.ft.)
80,000	550	44,000	660 m^2 (7104 sw.ft.)
100,000	500	50,000	750 m^2 (8073 sq.ft.)

within the area, accommodation for staff counters and catalogues, informal seating without table space on the scale of one seat per 1,000 population and a moderate amount of display equipment. If access to any part of the book-stack is permitted, this should not result in any reduction in the number of volumes accommodated on freely accessible open shelves on the scale indicated above.

Additional space requirements relevant to adult lending areas are discussed in the following paragraphs:

94–100.	Audio-visual materials
109.	Exhibition space.

Adult reference areas

88. Four different uses of space can be identified:

i Accommodation of open-shelf stock.
ii Enquiries desks and space for enquirers.
iii Accommodation for readers studying on library premises, and for users of 'quick reference' facilities.
iv Accommodation and consultation of periodicals.

Accommodation of open-shelf reference stock

89. The amount of open-shelf reference stock to be provided is discussed in paragraph 28. It may need to vary from a basic collection of 100 titles or so in the smallest service point (where it constitutes little more than 1% of the total stock of 9,000 volumes) to a total of perhaps 10% of the working stock in large libraries, where a minimum total working stock amounting to two volumes per head of population is recommended. The proportion of the total working stock which is allocated to adult reference use will tend to increase with the size of the population served. Other factors however influence the extent of the reference provision in each library. They include the extent to which the community is self-contained and hence dependant upon its own resources, the range of local interests and industry, the relations which have developed between the public library and other libraries in the vicinity, and the length of time during which the stock has been built up.

90. Although reference books may require on average more shelf space per volume than books for loan, space for the circulation of readers in open-shelf areas does not need to be so generous as that described for lending areas in paragraph 87, and it has been considered sufficient to allow 10 m^2 (108 sq.ft.)

per 1,000 volumes on open shelves. Table 2 shows, for libraries serving populations of different sizes, the open-shelf reference stocks which are considered likely in most cases to be necessary and adequate as basic working collections, together with the space required for their accommodation, and the additional areas likely to be needed for seating. (See paragraph 91). Together, these will allow sufficient space to accommodate any staff desks which are needed in the public area.

Table 2 Adult reference facilities

Population served	Minimum total working stock of the library as recommended in para. 22.		Open-shelf bookstock (Paragraphs 89–90)		Seating accommodation (Paragraph 91)	
	Per head of pop.	Total number of volumes	Number of volumes (Percentage of total stock in brackets)	Area required at 10 m^2 per 1,000 volumes	Number of places at 1.5 per 1,000 pop.	Area required at 2.5 m^2 per place
3,000	3	9,000	100 (1%)	1 m^2 (11 sq.ft.)	5	13 m^2; (140 sq.ft.)
5,000	3	15,000	300 (2%)	3 m^2 (32 sq.ft.)	8	20 m^2 (215 sq.ft.)
10,000	3	30,000	900 (3%)	9 m^2 (97 sp.ft.)	15	38m^2 (410 sq. ft.)
20,000	3	60,000	3,000 (5%)	30 m^2 (323 sq.ft.)	30	75 m^2 (807 sq.ft.)
40,000	2 1/2	100,000	7,000 (7%)	70 m^2 (753 sq.ft.)	60	150 m^2 (1614 sq.ft.)
60,000	2	120,000	12,000 (10%)	120 m^2 (1292 sq.ft.)	75	188m^2 (2024 sq.ft.)
80,000	2	160,000	16,000 (10%)	160 m^2 (1722 sq.ft.)	120	300 m^2 (3228 sq.ft.)
100,000	2	200,000	20,000 (10%)	200 m^2 (2153 sq.ft.)	150	375 m^2 (4035 sq.ft.)

Note: Space required for the housing and consultation of periodicals (see paragraph 92) and the storage and use of audio-visual materials (see paragraphs 93–100) must be added to the areas noted in the preceding table.

Accommodation for users studying on library premises, and for users of 'Quick Reference' facilities

91. The level of provision needed here will be influenced by many factors, including:

The nature of the population, including density, age structure, and the number of full-time students.

Home circumstances in the area.

The strength of the library's collections, including the provision of research resources and audio-visual materials.

The accessibility of other libraries.

The adequacy of academic libraries in the vicinity, and their policy with regard to the admission of readers.

A satisfactory level of provision in most circumstances would be 1.5 seats (each with table space) per 1,000 population, but it may sometimes be possible to reduce this level of provision in libraries serving populations larger than about 100,000. Allow about 2.5 m^2 (27 sq.ft.) per reader space, depending upon the arrangement of seats and tables. No purpose-built library should provide fewer than 4 seats with table space, suitable for study.

Housing and consultation of periodicals

92. In all libraries there is a need to provide some newspapers and periodicals, and accommodation for readers wishing to consult them. The scale of provision discussed in paragraphs 39–40 will be affected by factors similar to those noted in the previous paragraph. The number of seats needed for readers consulting periodicals will probably be at least one per 2,000 population in libraries serving up to about 20,000 population, but could well be reduced to one seat per 3,000 population above this figure. It is necessary, however, to emphasize the wide variation in need likely to result from different local circumstances. 3 m^2 (32 sq.ft.) of floor space should be provided per seat. This will provide sufficient space to display the current issues of newspapers and periodicals, and allow for the creation of a comfortable reading lounge. Upholstered chairs and low tables may be more appropriate than formal study tables, and they can sometimes be associated with the informal seating in the adult lending library (see paragraph 87) to create an all-purpose lounge and browsing area. No purpose-built library should provide fewer than eight informal seats for all purposes. The more scholarly periodicals will commonly be consulted in the reference library rather than

in this informal area, and many of them will be more suitably housed there. When this type of use constitutes a considerable proportion of the total usage of periodicals, it may be desirable to allocate additional seating accommodation in the reference library and reduce the informal seating accordingly.

Audio-visual materials and equipment

93. Paragraphs 42 to 47 discuss the nature of audio-visual materials and note that their provision in public libraries is at present undergoing rapid change and expansion. In these circumstances accommodation standards cannot easily be formulated or recommended with confidence. A number of ways in which the provision of audio-visual materials affects the planning of library buildings must however be noted.

The storage of audio-visual materials

94. In large libraries it will usually be most satisfactory that audio-visual materials should be accommodated with the departmental collections to which they most nearly relate. Some audio-visual materials, however, should be stored if possible at an even temperature and in a dust-free environment, and magnetic recordings must be kept away from magnetic fields such as those produced by heavy wiring installations. These considerations may make it necessary, or convenient, even in a large departmentalized library, to bring together much of the collection of audio-visual materials into a central store, preferably closely associated with the central store for audio-visual equipment (see paragraph 96). The planning of new libraries should ensure that any such central collection is easily accessible from each of the areas or departments with an interest in it.

95. The space required for accommodating audio-visual materials will increase with the size of the library, and the extent to which collections have been built up. Gramophone records can be expected to make the greatest demands, and open access collections of records, displayed in browsing boxes, will require generous space allocation.

The storage of audio-visual equipment

96. Accommodation must also be provided in the library for a considerable and growing range of equipment needed in order to make use of audio-visual

materials. Some of this equipment, such as gramophone turntables and microfiche readers may sometimes be kept permanently set up and ready for use. Other items such as tape recorders and slide projectors, will often be stored on shelves until needed, and accommodation for equipment of this kind will usually be centralized within the library, and sometimes even within the administrative unit. Large libraries will employ a technician to look after audio-visual equipment, and will also wish to install other photographic and recording equipment so that they may extend the range of material available to them.

97. Any library serving a population of 20,000 or more will probably need to provide separate storage accommodation for audio-visual equipment. In libraries serving populations of 60,000 or more it will often be necessary to provide an additional area to accommodate a technician, laboratory and other photographic and recording equipment.

The use of audio-visual materials and equipment

98. The planning of a library with a significant multimedia collection should provide for the use of these materials on library premises both by individuals and by groups. There will be particular value in providing for the use of audio-visual materials in children's libraries. The needs of groups will be met by the provision of meeting rooms with suitable viewing and listening facilities, and the provision of such rooms, adequately equipped for the use of audi-visual materials, therefore becomes more important as library collections of these materials increase.

99. The needs of individual users do not necessarily result in additional space requirements. The objective of library planning should be to make possible the use of audio-visual material at any study desk within the library, and although this ideal may not always be achieved there must at least be generous provision of electrical power outlets to study desks and to any individual study room provided in reference areas. It may sometimes be found that the provision of audio-visual facilities in this way will bring additional users to the library, and justify a more generous provision of place than that suggested in paragraph 91. Although sound recordings, radio, etc., will usually be listened to through earphones, some sound-proofing of individual and group study rooms may be essential. Full black-out is not necessary for individual viewing on efficient equipment, but the layout of the library must

make it possible to view slides, microforms, video tapes, and other projected materials without the distraction of direct sunlight.

100. The only increase in space requirements which is envisaged in connection with the use by individuals of audio-visual materials will occur when equipment such as gramophone turntables or microform readers is kept permanently installed and ready for use by either staff or members of the public, or when seats provided with jack sockets are set aside for the sole purpose of listening to records. For each installation of this kind an additional 2.5 m^2 (27 sq.ft.) should be provided.

Library services for children

General

101. Four main needs can be identified:

i Provision of lending facilities, including accommodation for books, other materials and displays.

ii Provision of reference materials and of study space, including accommodation for children doing school homework.

iii Accommodation for individual viewing and listening.

iv Accommodation for library-sponsored activities such as story hours, filmshows, talks and demonstrations.

It will usually be possible to meet the first three of these needs within a single children's area, which itself will often be only a part of an open-plan general area, at least in small libraries. Provision for activities of various kinds is discussed in paragraph 108.

102. Children's libraries are peculiarly prone to peaks of demand, especially during the hour or two after schools close. Space must be related to these periods of peak demand. Another important planning consideration is that children's libraries are commonly visited by school classes during the day, and must be able to accommodate such visits without disrupting the general service of the library.

103. Young children, particularly, are seldom able to make long journeys unaccompanied to visit the library, and for this reason the population effectively served by a children's library may sometimes be less than that served by adult departments within the same building. However, there is evidence of an increase in family visiting to the library, and where this occurs it has important implications for the planning of library buildings. It can no longer

be assumed that main and large branch libraries will serve only the children in their immediate vicinities, and bookstock, accommodation and other facilities may have to be enlarged accordingly. This, however, will not remove the need to provide additional service points readily accessible to children. Children and adults should not be required to enter library buildings by separate doors, and there should be no obstacles to communication between parents and children once inside.

104. There may be a few main libraries in large cities which have little need to provide children's facilities, being situated in districts given over entirely to commerce. Elsewhere, there may be a need to provide children's libraries in separate buildings in areas where there are many children and from which they cannot easily reach a general library. Separate facilities for adults and children, however, should be provided in this way only when necessitated by circumstances.

Lending areas

105. Notions as to the book capacity needed vary considerably from country to country. It must be sufficient to display the full variety of books needed to meet the demands of users, many volumes – particularly those for young children – being displayed face-forward for maximum effect. Allow 16 m^2 (172 sq.ft.) for every 1,000 volumes on open shelves. This assumes that shelf units for children will be four shelves high. Provision on this scale will allow – as in the adult lending area – for circulation of readers, accommodation for staff counters and catalogues, informal seating without table space and a moderate amount of display equipment.

106. In libraries serving a total population up to 10,000 the area needed for the above purposes will commonly be 75 to 100 m^2 (807 to 1076 sq.ft.). Between 10,000 and 20,000 population, an area of 100 to 200 m^2 (1076 to 2152 sq.ft.) will usually be needed. Libraries serving larger populations will need more accommodation than this, depending on local patterns of use. In all cases, there should be sufficient space to accommodate a class of schoolchildren without seriously disturbing other users.

Study space

107. Space for a few tables and chairs for casual use should be added to the children's lending area. Many older children who need accommodation for

homework and serious study will prefer – and need – to use the adult reference library, and should not be discouraged from doing so. Space for periodicals, however, should also be provided in the children's library. Sometimes there will be particular local circumstances which make it desirable to provide additional reference accommodation for children in the children's library.

Accommodation for activities

108. Activities such as talks and story hours are a common feature of library service to children. They can often be carried out in part of a well-planned children's area without interfering with the work of the library as a whole.

Other activities such as dramatic performances and filmshows will need the projection, black-out and other facilities available in a general-purpose meeting room, which should therefore be conveniently accessible from the children's area. There may still be a need, however, to set aside a specially designed room as an annex to the children's library, if justified by the extent, or by the nature, of children's activities. If it is to be used for activities in which children are essentially the audience, it will be sufficient to allow 1.5 m^2 (16 sq.ft.) per place. There might be more value, however, in developing a club-room atmosphere in which the children can participate in a range of creative activities, materials for which can be accumulated and used in the room without the need to tidy everything away after each session. For such purpose as this an area of up to 3 m^2 (32 sq.ft.) per place will be needed, A children's room of this kind should be as attractive and informal as possible, and should provide accommodation for at least 30 children(or a school class) and possibly up to 100 according to local circumstances.

Exhibition space

109. Every public library should provide opportunities for exhibitions, not only of books but of other objects and illustrative material of various kinds, as an extension of its educational, cultural and informational functions. It is better that exhibition space should be integrated into the public library areas than that any specific room or separated area should be set aside for the purpose, though problems of security, lighting and humidity control may sometimes restrict the nature of the material which can be exhibited in this

way. In order to accommodate exhibitions up to 10% should be added to the public area in one or more of the main library departments. With careful planning space allowed for exhibition purposes will often be capable of alternative uses, and will provide further opportunities for flexibility within the building. Provision must be made for the preparation and storage of exhibition materials, and it will usually be convenient to set aside a special room, easily accessible from delivery vehicles, for the purpose.

Storage of reserve stock

110. The extent to which it is necessary to provide book storage accommodation in any particular library building will depend on the size and nature of reserve stocks and policy regarding their housing. Provision should be made, where necessary, for estimated future increases in the need for book storage capacity. A choice must be made between closed access stacks and stacks with some degree of open access. In the former, provide 5.5 m^2 (59 sq.ft.) per 1,000 volumes (equivalent to 182 volumes per m^2), but this capacity will be approximately doubled if compact shelving is employed. In stacks to which the public have access, provide 7 m^2 (75 sq.ft.) per 1.000 volumes (equivalent to 143 volumes per m^2). Open access stacks should be readily accessible from public areas.

Staff workrooms and offices

111. Even the smallest library requires some working accommodation for staff outside the public areas. Nevertheless it is important that some staff work areas should be associated with, or easily accessible from, each of the areas of public service. In libraries the working accommodation for staff combines the functions of office, factory and storeroom. Space must be provided not only for people seated at desks, but also for work benches, equipment and materials, including the constant ebbing and flowing of books. The consequence is that normal standards of office accommodation have very little relevance to the needs of libraries, which are affected by the number of people likely to be in the work areas at any one time, by the way in which the library system is organized and by the complexity of the operations required to be undertaken at each individual service point.

112. These circumstances make it impossible to recommend a simple and generally applicable formula for the calculation of the areas of library work-

rooms and offices. But they also suggest that there will usually be a close relationship between the workroom and office requirements of a library and the areas of the main public departments. 20% added to the total area of public departments will usually be found to provide adequately for workrooms and offices, and will be equivalent to about 10–12 m^2 (108–129 sq.ft.) of office or workroom space per staff member.

113. Administrative offices for a group of libraries (for example, in a main library or large branch library) will require additional accommodation, and this is likely to be proportional to the size of the population served. For such purposes allow an additional 20 m^2 (215 sq.ft.) per 10,000 population.

Staff rest rooms, kitchen, toilets, cloakroom, etc

114. All libraries should provide at least basic staff amenities, but in very small branch libraries these may not always be in separate and specialized rooms but may in some cases be combined with workrooms and storage accommodation. For each member of staff who may be expected to use the facilities (with generous allowance for probable future increase in staff) allow 2–4 m^2 (22–43 sq.ft.) on a scale varying inversely with the size of the staff:

Number of staff	Area per staff member	Total area
2	4.0 m^2 (43 sq.ft.)	8 m^2 (86 sq.ft.)
10	4.0 m^2 (43 sq.ft.)	40 m^2 (431 sq.ft.)
20	3.0 m^2 (32 sq.ft.)	60 m^2 (646 sq.ft.)
50	2,4 m^2 (26 sq.ft.)	120 m^2 (1292 sq.ft.)
100	2.2 m^2 (24 sq.ft.)	220 m^2 (2368 sq.ft.)
200	2.0 m^2 (22 sq.ft.)	400 m^2 (4306 sq.ft.)

Circulation space or 'balance area'

115. Circulation space consists of entrance halls, stairwells, lavatories, cloakrooms, lobbies and corridors outside the areas whose accommodation requirements have been separately discussed. In public parts of the building circulation is effected largely within the specified areas, and it is desirable that corridors and other circulation spaces should be kept to a minimum. In staff areas proportionately more circulation space will be required.

116. Allow 10–15% of all public areas and 20–25% of all staff areas, the higher figure in each case being unlikely to be necessary except in large

libraries with a high degree of division into separate rooms and departments. The lower figures given above – 10% and 20% respectively – should be regarded as 'balance' area, to provide circulation space as described above, but also available for any desired purpose within the building if careful planning has made it unnecessary to provide circulation space on this scale.

Other areas

117. Many libraries need to provide areas for specialized purposes not covered by these recommendations. They include the administration of additional services such as those to schools, hospitals and the housebound; bases for mobile libraries; archives depositories; public facilities such as refreshment areas and meeting rooms, and services for the building such as heating and ventilating plant and general storage accommodation, especially for cleaning materials and equipment; all with their related circulation space. These needs vary considerably from library to library, and standards for their provision cannot usefully be recommended. All must be taken into consideration, however, when a library is planned, and they will result in a need for accommodation additional to that already discussed and provided for.

COST OF THE PUBLIC LIBRARY SERVICE

118. International standards of cost are not possible, since the major items of salaries, materials, and accommodation vary widely from country to country. But it is generally observable that in the small unit, the cost of materials tends to be high in proportion to the total budget, and in the larger administrative unit, the proportion spent on salaries tends to increase.

The provision of new services, such as audio-visual materials, development of the public library as a cultural centre, and facilities for handicapped readers, will also increase costs. Even so the cost of the public library service, in terms of the value it gives to the community, will still be small, even at its most developed.

119. If the levels of provision recommended in this document are achieved, a public library system can expect to improve the range and quality of the service it gives, with a consequent increase in the volume of use and in the value of its service to the community.

International Federation of Library Associations and Institutions

Serial IFLA Publications

Edited by Willem R. H. Koops and Peter Havard-Williams

1 **Special Libraries – Worldwide.** A collection of papers prepared for the Section of Special Libraries. Edited by Günter Reichardt. 1974. 360 pages. DM 68.00, IFLA members DM 51.00. ISBN 3-7940-4421-5

2 **National Library Buildings.** Proceedings of a colloquium held in Rome, 3–6 September 1973. Edited by Anthony Thompson. 1975. 144 pages. DM 28.00, IFLA members DM 21.00. ISBN 3-7940-4422-3

3 **Le contrôle bibliographique universel dans les pays en développement.** Table ronde sur le contrôle bibliographique universel dans les pays en développement, Grenoble, 22–25 août 1973. Edité par Marie-Louise Bossuat, Geneviève Feuillebois, Monique Pelletier. 1975. 165 pages. DM 38.00, IFLA members DM 29.00. ISBN 3-7940-4423-1

4 **National and International Library Planning.** Key papers presented at the 40th Session of the IFLA General Council, Washington, DC, 1974. Edited by Robert Vosper and Leone I. Newkirk. 1976. 162 pages. DM 36.00, IFLA members DM 27.00. ISBN 3-7940-4424-X

5 **Reading in a Changing World.** Papers presented at the 38th Session of the IFLA General Council, Budapest, 1972. Edited by Foster E. Mohrhardt. 1976. 134 pages. DM 28.00, IFLA members DM 21.00. ISBN 3-7940-4425-8

6 **The Organization of the Library Profession.** A Symposium based on contributions to the 37th Session of the IFLA General Council, Liverpool 1971. Edited by A. H. Chaplin. 1976. 2nd edition. 132 pages. DM 28.00, IFLA members DM 21.00. ISBN 3-7940-4300-X

7 **World Directory of Administrative Libraries.** A guide of libraries serving national, state, provincial, and Länder-bodies, prepared for the Sub-Section of Administrative Libraries. Edited by Otto Simmler. 1976. 474 pages. DM 60.00, IFLA members DM 45.00. ISBN 3-7940-4427-4

8 **World Directory of Map Collections.** Compiled by the Geography and Map Libraries Sub-Section. Edited by Walter W. Ristow. 1976. 326 pages. DM 48.00, IFLA members DM 36.00. ISBN 3-7940-4428-2

9 **Standards for Public Libraries.** 1977. 2nd corrected edition. 53 pages. DM 12.80, IFLA members DM 9.80. ISBN 3-7940-4429-0

10 **IFLA's First Fifty Years.** Achievement and challenge in international librarianship. Edited by Willem R.H. Koops and Joachim Wieder. 1977. 158 pages. DM 36.00, IFLA members DM 27.00. ISBN 3-7940-4430-4

11 **The International Federation of Library Associations and Institutions. A Selected List of References.** 2nd edition, revised and expanded. Compiled by Edward P. Cambio. 1977. VI, 52 pages. DM 16.80, IFLA members DM 12.80. ISBN 3-7940-4431-2

12 **Library Service to Children.** Edited by Colin Ray. 1977. ISBN 3-7940-4423-0. In preparation for 1978

Verlag Dokumentation, Publishers – POB 711009 – D-8000 München 71

FSC
www.fsc.org
MIX
Papier | Fördert gute Waldnutzung
FSC® C083411

Kolibri 360 GmbH
Ferdinand-Jühlke-Straße 7
99095 Erfurt, Deutschland
produktsicherheit@kolibri360.de